Anonymous

Buffalo Lithia Springs, Mecklenburg County, Virginia

Anonymous

Buffalo Lithia Springs, Mecklenburg County, Virginia

ISBN/EAN: 9783337817039

Printed in Europe, USA, Canada, Australia, Japan

Cover: Foto ©berggeist007 / pixelio.de

More available books at **www.hansebooks.com**

Dr. J. MARION SIMS, of New York.

"I HAVE USED IN MY PRACTICE THE BUFFALO LITHIA WATER, SPRING No. 2, FOR TWO YEARS PAST, AND HAVE, IN MANY CASES, FOUND IT HIGHLY EFFICACIOUS."

BUFFALO LITHIA SPRINGS,
MECKLENBURG COUNTY, VIRGINIA.

HEALTH PRIMER.

STONE IN THE BLADDER

(In which this Water is the Only Known Solvent),

BRIGHT'S DISEASE OF THE KIDNEYS, GOUT RHEUMATIC GOUT, RHEUMATISM, DISORDERS OF THE, STOMACH AND NERVOUS SYSTEM, AFFECTIONS PECULIAR TO WOMEN, ECZEMA, ACNE, AND OTHER FACIAL ERUPTIONS.

RESIDENT PHYSICIAN:

G. HALSTEAD BOYLAND, M.D.,

LATE PROFESSOR SURGERY BALTIMORE MEDICAL COLLEGE, LATE SURGEON FRENCH ARMY (DECORATED), MEMBER OF BALTIMORE ACADEMY OF MEDICINE, MEMBER OF THE MEDICAL AND CHIRURGICAL FACULTY OF THE STATE OF MARYLAND, &c., &c.

MEDICAL OPINIONS—CLINICAL REPORTS.

THOMAS F. GOODE, Proprietor,
POST-OFFICE ADDRESS:
BUFFALO LITHIA SPRINGS, VA.

PHILADELPHIA:
McCalla & Stavely, Printers, 237-9 Dock Street.
1884.

This Water, fresh from the Spring, is without Taste or Odor to distinguish it from ordinary drinking water.

LOCATION, TOPOGRAPHY, ACCOMMODATIONS, ANALYSES, &c.

BUFFALO LITHIA WATER

NATURE'S MATERIA MEDICA

TRADE MARK—PATENTED.

This watering place is situated in Mecklenburg Co., Virginia, 500 feet above the level of the sea, and is centrally located in a section of country about eight miles square, known as the "Buffalo Hills," which is broken and rolling, and very similar in general appearance to that lying just under the Blue Ridge of mountains known as "Piedmont Virginia." It is twelve miles distant from the Scottsburg Depot of the Richmond and Danville Railroad and seven miles from the Clarksville Depot of the Richmond and Mecklenburg R.R., at either of which points, visitors for the Springs leave the cars. Passengers taken promptly from all trains.

There are comfortable accommodations for three hundred persons.

The Springs are three in number, known as Nos. 1, 2, and 3. The remedial value of all of them is a well-established medical fact.

Spring No. 2, however, is that now attracting so large a share of public attention, and to that Spring the testimonials embraced in this pamphlet for the most part have reference.

ANALYSES.

Analyses of the Waters of the Buffalo Springs. in Mecklenburg county, Virginia, made by Prof. Wm. P. Tonry, of the Maryland Institute, Baltimore, March 17th, 1874. (Results expressed in grains per imperial gallon.)

SPRING No. 1.	GRAINS.	SPRING No. 2.	GRAINS.
Sulphate of Magnesia	1.530	Sulphate of Magnesia	0.885
Alumina	8.180	Alumina	9.067
Potash	0.463	Lime	33.067
Lime	19.251	Carbonate of Potash	29.300
Bicarbonate of Lime	39.277	Bicarbonate of Lime	14.963
Lithia	1.484	Lithia	2.250
Iron	0.500	Baryta	1.750
Chloride of Sodium	1.256	Iron	0.300
Silica	1.725	Chloride of Sodium	4.921
Phosphoric acid	traces	Silica	1.873
Iodine	traces	Phosphoric Acid	traces
Organic matter	small amount	Iodine	traces
		Organic matter	small amount

Total number of grains per
gallon 73.693
Sulphuretted hydrogen ....5.9 cubic in.
Carbonic acid gas.......69.1   "

Total number of grains per
gallon 98.376
Sulphuretted hydrogen... .8.3 cubic in.
Carbonic acid gas........59.2   "

SPRING No. 3.

	GRAINS.		GRAINS.
Sulphate of Magnesia	0.150	Chloride of Sodium	0.217
Alumina	3.035	Silica	0.570
Lime	2.353	Phosphoric Acid	traces
Carbonate of Potash	1.852	Organic matter	small amount
Bicarbonate of Lime	2.524		
Lithia..traces of Lithia			
Iron	3.774		

Total number of grains per
gallon 14.475
Sulphretted hydrogen.....3.4 cubic in.
Carbonic acid gas........11,6   "

NATURAL MINERAL WATER,

Depots: { 112 N. Ninth Street, Philadelphia, Pa.
47 and 49 N. Charles St., Baltimore, Maryland,
1010 F. Street, Washington, D. C.

ALWAYS IN STOCK.

Domestic Waters.

ALLEGHANY,	Virginia.
BATH ALUM,	Virginia.
BEDFORD, (Alum and Iron,)	Virginia.
BEDFORD,	Pennsylvania.
BUFFALO LITHIA,	Virginia.
BLUE RIDGE,	Virginia.
BLUE LICK, (Sulphur,)	Kentucky.
BETHESDA, CLYSMIC, SILURIAN, }	Waukesha, Wisconsin.
CAPON,	West Virginia.
DEEP ROCK,	Oswego, New York.
FARMVILLE, LITHIA,	Virginia.
MOUNT CLEMENS,	Michigan.
MASSANETTA,	Virginia.
PANACEA,	North Carolina.
ROCKBRIDGE ALUM,	Virginia.
ROCK ENON, (Iron,)	Virginia.
STRONTIA,	Maryland.
TAYLOR, (now Massanetta,)	Virginia.
WHITE SULPHUR,	West Virginia.
RED SULPHUR,	West Virginia.
RAWLEY, Iron)	Virginia.
WALLAWHATOOLA, (Alum,)	Virginia.

Saratoga Waters.

CHAMPION,	Geyser.
HATHORN,	Empire.
CONGRESS,	Vichy.

Foreign Waters.

FRIEDRICHSHALL,	From Germany.
RHENS' (Seltzers,)	From Germany.
HUNYADI JANOS,	From Germany.
OFNER RAKOCZY,	From Hungary.
KISSINGEN, (Rakoczy)	From Bavaria.
CARLSBAD, { Sprudel Muhlbrunn }	from Bohemia
PULLNA,	From Bohemia.
APOLLINARIS,	Rhenish.
VICHY, { Celestins Grand Grille }	From France.

Syphoned Waters.

NATURAL.

PLAIN SODA.

CHAMPION,	Saratoga.
GEYSER,	Saratoga.
VICHY,	Saratoga.
OSWEGO DEEP ROCK.	
BEDFORD, OF PENNSYLVANIA.	
STRONTIA, OF MARYLAND.	
SILURIAN, OF WISCONSIN.	
MASSANETTA, OF VIRGINIA.	

GENERAL TESTIMONIALS.

Affections of the Nervous System, Complicated with Bright's Disease of the Kidneys, or with a Gouty Diathesis. Cerebral Hyperæmia, &c.

Dr. Wm. A. Hammond, of New York, Surgeon-General U. S. Army (retired), Professor of Diseases of the Mind and Nervous System in the University of New York, &c.

" I have for some time made use of the Buffalo Lithia Water in cases of affections of the *Nervous System* complicated with *Bright's Disease of the Kidneys*, or with a *Gouty Diathesis. The results have been eminently satisfactory.* Lithia has for many years been a favorite remedy with me in like cases, but the *Buffalo Water certainly acts better than any extemporaneous solution of the Lithia Salts, and is, moreover, better borne by the stomach.* I also often prescribe it in those cases of *Cerebral Hyperæmia, resulting from over-mental work*—in which the condition called *Nervous Dyspepsia* exists—*and generally with marked benefit.*"

Chronic Interstitial Nephritis, or Bright's Disease of the Kidneys, Gouty and Rheumatic Affections, &c.

Dr. Alfred L. Loomis, of New York, Professor of Institutes and Practice of Medicine, Medical Department University of New York, Visiting Physician Bellevue Hospital, Consulting Physician Charity Hospital, New York.

" For the past four years I have used the Buffalo Lithia Water in the treatment of Chronic *Interstitial Nephritis** occurring in *Gouty* and *Rheumatic* subjects, with the most *marked benefit.* In all *Gouty* and *Rheumatic Affections*, I regard it as *highly efficacious.*"

---

* Ziemssen in his great work, "The Cyclopædia of the Practice of Medicine," under the head of "*Interstitial Inflammation of the Kidneys*" (that is *Interstitial Nephritis*), says, "The pathological state of the *Kidneys* at present designated by the above name (along with several other names), represents *the third stage* of what is known by authors as *Bright's Disease.*

Gravel and other Diseases dependent upon a Uric Acid Diathesis, Rheumatic Gout, Dyspepsia, Diseases Peculiar to Women, &c. This Water a Powerful Nervous Tonic.

Dr. Hunter McGuire, Richmond, Virginia, late Professor Surgery Medical College of Virginia.

" Buffalo Lithia Water, Spring No. 2, as an *Alkaline Diuretic*, is invaluable. In *Uric Acid Gravel*, and indeed in diseases generally dependent upon a *Uric Acid Diathesis*, it is a remedy of extraordinary potency. I have prescribed it in cases of *Rheumatic Gout*, which had resisted the ordinary remedies, with wonderfully good results. I have used it also in my own case, being a great sufferer from this malady, *and have derived more benefit from it than from any other remedy.* It has very marked adaptation in Diseases of the *Digestive Organs.* In that condition, especially known as *Nervous Dyspepsia*, frequently caused by over-mental labor, and in those cases also where there is *excess of Acid* in the process of nutrition, it will be found highly efficacious.

" I am fully satisfied of its value in the treatment of *Diseases Peculiar to Women.* In this class of disease *it is unquestionably deserving of very high commendation.* It has never failed me as a *Powerful Nervous Tonic* when I have prescribed it as such. I sometimes think it must contain *Hypophosphites of Lime and Soda. It acts as that compound does—as a Tonic and Alterative.*"

This Water Compared with that of the Celebrated White Sulphur Springs in Greenbrier County, West Virginia. Its value in Malarious Fevers, Affections Peculiar to Women, Atonic Dyspepsia, &c.

Dr. Wm. T. Howard, of Baltimore, Professor of Diseases of Women and Children in the University of Maryland.

Dr. H. attests the *common adaptation of this Water in " a wide range of cases,"* with that of the far-famed White Sulphur Springs, in Greenbrier county, West Virginia, and adds the following :

"Indeed, in a certain class of cases, it is much superior to the latter. I allude to the abiding debility attendant upon the tardy convalescence from grave acute diseases; and more especially to the *Cachexia* and *Sequels* incident to *Malarious Fevers*, in all their grades and varieties, to certain forms of *Atonic Dyspepsia*, and *all* the *Affections Peculiar to Women* that are remediable at all by mineral waters. *In short, were I called upon to state from what mineral waters I have seen the greatest and most unmistakable amount of good accrue in the largest number of cases in a general way, I would unhesitatingly say the Buffalo Springs, in Mecklenburg county, Virginia.*"

Solvent Power of Spring No. 2 in Uric Acid and Phosphatic Deposits, the Gouty Diathesis, Albuminuria, Eczema, Acne and other Facial Eruptions.
Springs Nos. 1 and 3.

Dr. G. Halstead Boyland, Late Professor of Surgery, Baltimore Medical College, Late Surgeon French Army (Decorated), &c., &c.

" I have made frequent and free use of the Buffalo Lithia Waters in my practice. In *Stone* in the *Bladder* of the *Red Lithic Acid*, and the *White Phosphatic Deposit*, the *Solvent* power of Spring ' No. 2 ' is unmistakable. The best results which I have witnessed from any remedy in *Gout* have been from this water, in which its solvent action upon the *Uric Acid* Deposit is equally evident. Its value, however, in such cases is by no means limited to its *solvent* power over these deposits, *but it meets the more important indication, that of so changing the Diathesis, on which the formation depends, as to prevent re-formation.*

By repeated chemical analyses, I have demonstrated the unquestioned power of this water in *Albuminuria.* Its *Nerve Tonic* properties are very decided, indicating its value in a wide range of *Nervous* disorders. In *Acid Dyspepsia. Cardiac Dropsy,* and in *Constipation* due to *Anæmia.* I have found it a remedy of great efficiency. In *Eczema, Acne,* and other *Facial Eruptions* its action is sometimes surprisingly happy.

"In *Debility* caused by *Uterine* displacement, the water of Spring ' No. 1 ' has almost a specific action upon the external and internal organs of generation, especially in cases of *Flexion* and *Version*, where the *Uterine Nerves* respond well to its stimulating effect, bringing about muscular tone from want of which displacement occurs. The power of this water as an *Aphrodisiac* is very marked. In common with ' No. 2 ' it possesses *Nerve Tonic* properties, and indeed in a much greater degree.

" In cases of *Nervous Prostration* its restorative action is often very prompt, and, not unfrequently, very astonishing.

" It is a more potent invigorator of the *Appetite* and of the powers of *Digestion*, **and** in many forms of *Atonic Dyspepsia* will be found a remedy of greater efficiency. Spring 'No. 3' is a potent *Chalybeate* with decided *laxative* properties, *and will be found valuable in many cases in which Chalybeate waters generally are inadmissible because of their astringent properties.*"

Inflammation of the Uterus and Bladder, Gouty, Rheumatic or Acid Diathesis, Neuralgia and Dyspepsia, Scarlet Fever, Albuminuria of Pregnancy.

Dr. James B. McCaw, Professor of the Practice of Medicine in the Virginia Medical College.

Extract from the Proceedings of the Richmond Academy of Medicine, October 15th, 1878, taken from the Virginia Medical Monthly, of December, 1878.

" *Buffalo Lithia Water of Mecklenburg County, Va., in Female and other Diseases.*—Prof. James B. McCaw (*Professor of the Practice of Medicine* in the Virginia Medical College at Richmond), reported two cases of great irritability of the *Uterus* and *Bladder when all other treatment had failed—both local and general* —which were very much relieved by the use of the ' Buffalo Lithia Water.' "

Dr. McCaw also spoke of the great value of these waters in the " *Gouty Rheumatic* or *Acid Diathesis,*" in " *Neuralgias* and *Dyspepsias,*" in the management of " *Scarlet Fever,*" and in "*Albuminuria of Pregnant Women.*"

The Gouty or Uric Acid Diathesis Affections Peculiar to Women, Acid Dyspepsia, Bright's Disease, &c. This Water a Nervous Tonic and Exhilarant.

Dr. Harvey L. Byrd, of Baltimore, President and Professor of Obstetrics and Diseases of Women and Children, in the Baltimore Medical College, formerly Professor of Practical Medicine, &c.

" I have witnessed the best results from the action of the Buffalo Lithia Water, Spring No. 2, in *Chronic Gout, Rheumatic Gout, Rheumatism, Gravel* and *Stone* in the *Bladder,* and I do not hesitate to express the opinion that in *all* diseases depending upon or having their origin in *Uric Acid Diathesis,* it is *unsurpassed, if, indeed, it is equaled by any water thus far known to the profession.*

" In some of the *Affections Peculiar to Women,* in the sub-acute and chronic form, especially in *Leucorrhœa, Amenorrhœa, Dysmenorrhœa,* and *Cystirrhœa,* I have in numerous cases found it highly efficacious. In *Cystirrhœa* it may be regarded as *well nigh specific.*

" I have prescribed this Water with the most satisfactory results, both as a *remedy* and *prophylactic* in the *Parturient* or *Pregnant* condition, for the relief of troublesome *vomiting* and the prevention of *Puerperal Eclampsia* or *Convulsions;* and I may say that I know of no remedy of equal efficacy with the Water of Spring No. 2 in the *Sequelæ of Scarlatina.*

" It is an admirable general *Tonic* and *Restorative, increasing the Appetite, promoting Digestion,* and *Invigorating the General Health.* It is *powerfully Antacid,* and especially efficacious in what is commonly known as *Acid Dyspepsia.* It is strongly commended to a very large class of sufferers by a peculiar power as a *Nervous Tonic* and *Exhilarant,* which makes it exceedingly valuable, where there is nothing to contra-indicate its use, in all cases where *Nervous Depression* is a symptom.

" It has an *ascertained value* in *Bright's Disease. A knowledge of its action in that disease thus far would seem* to warrant the belief that it would, in many instances, *at least in its early stages, arrest it entirely ; and in its more advanced stages prove a decided comfort and palliative.*"

The Gouty Diathesis, Chronic Inflammation of the Bladder, &c.

Dr. Alexander B. Mott, of New York, Professor Surgery Bellevue Hospital Medical College, Surgeon Bellevue Hospital.

"I have made sufficient use of the Buffalo Lithia Water to be satisfied that it possesses very valuable *therapeutic* properties. In the *Gouty Diathesis, Chronic Inflammation* of the *Bladder*, and other diseases affecting the *Urinary Organs*, it may be relied on to give the most satisfactory result."

Puerperal Eclampsia.

Dr. William Thompson Lusk, of New York, Professor of Obstetrics and Diseases of Women and Children, in the Bellevue Hospital Medical College, Fellow of the American Gynæcological Society, Author of "The Science and Art of Midwifery," &c.

Dr. Lusk in "The Science and Art of Midwifery," after speaking of the value of the "milk diet" in *Puerperal Eclampsia*, advises, that where the diet is badly supported by the patient, she should be advised to drink freely of the natural *Alkaline Waters*, possessing mildly diuretic properties, and suggests the "*Buffalo Lithia Water.*"

This Water a Powerful Agent for the Removal of Vesical Calculi. Its Value in Bright's Disease of the Kidneys, the Gouty Diathesis, Malarial Cachexia, &c.

The late Dr. Thomas P. Atkinson, formerly of Danville, Va., at one time President Medical Society of Virginia.

* * * * * * *

"Experience has proved the Buffalo Lithia Water to be a powerful agent for the removal of *Vesical Calculi.*

"It has been found especially efficacious in the *Uric Acid* variety.

"Its beneficial results, however, in *Uric Acid Calculi* are not restricted to the removal by means of solution or disintegration of *Calculi* which have been already deposited, but it not unfrequently so corrects the constitutional tendency to excess in the production of *Uric Acid*, or of any other acid which may have the property of precipitating it from its solution, as to prevent further deposition. It has in some cases proved an efficient remedy in effecting the solution and preventing the deposition of the *Phosphatic* as well as the *Uric Acid* sediment.

* * * * * * *

"When used at any early stage, while enough of the renal structure remains to answer the purpose of purifying the blood, it is of decided efficacy in *Bright's Disease of the Kidneys*, and indeed in some cases where the destruction of the *Kidney* has been greater, its use has resulted in partial restoration and prolongation of life.

* * * * * * *

"Its beneficent influence in *Gout*, and I may add in *Rheumatism*, is almost universal, embracing much of the so-called *Neuralgia*, now so prevalent, which is simply *Gout* or *Rheumatism* in the *Nervous Form.*

* * * * * * *

"In *Chronic Intermittent* and *Remittent Fevers*, and in the whole of that large class of disorders included in the term *Malarial Cachexia*, embracing *Enlargement* of the *Liver* and *Spleen, Jaundice, Indigestion* and *Neuralgias*, of which the malarious districts of the South are so prolific, it will be found uniformly and highly efficacious."

Albuminuria and Nausea of Pregnancy.

Dr. G. A. Foote, of *Warrenton, North Carolina, Ex-President State Medical Society, Member of the Board of State Medical Examiners, and Member of the State Board of Health.*

" I have frequently used the Buffalo Lithia Water (Spring No. 2) in the *Albuminuria* attending *Pregnancy*, and I have no hesitancy in saying that I regard it as the most *efficacious* remedy I ever prescribed for this malady. I have also prescribed it, and am now using it, in the *Nausea of Pregnancy*, with the most marked and gratifying results."

Uric Acid Diathesis, Renal Congestion and Uræmic Poisoning complicating Pregnancy.

Dr. Martin L. James, of *Richmond, Va., Professor of Materia Medica and Therapeutics, Medical Society of Virginia.*

[Proceedings of the Richmond, Virginia, Academy of Medicine, Dec. 16th, 1880.]

" The President of the Academy, Dr. M. L. James, reported three cases of marked *Uric Acid Diathesis* successfully treated by the water of Spring No. 2 of the Buffalo Lithia Springs of Virginia. In one of these patients there were, as results of this Diathesis, *Sandy Deposits* in the *Urine, Inflammation of the Kidneys, Irritability of the Bladder* and *Hæmaturia*, and finally the passage of a *Calculus* of the size of a cherry, formed on a blood-clot as a nucleus. In another case there were frequent attacks of *Nephritic Colic*, attended by *Sandy Depositions* in the Urine. After the free use of this water in these cases, no further manifestations of the disorder occurred.

" Dr. James further reported a case of *Congestion* of the *Kidneys* in a lady eight months advanced in *Pregnancy, attended by marked Œdema* both over the extremities and surface, and by *Uræmic Poisoning to such an extent as very seriously impaired the vision of the patient*, relieved by the free use of this water for three weeks.

" Other remedies, he stated, were used in these cases, but the favorable results seemed clearly attributable to the action of the water."

Gouty Diathesis.

Dr. Horatio C. Wood, *Professor of Materia Medica, &c., in the Medical Department of the University of Pennsylvania, in the " Medical Times " of July 20, 1878.*

" The water of the Buffalo Springs, of Mecklenburg county, Virginia, was brought to our notice by a Baltimore physician, who had been relieved by its use of some very troublesome and alarming symptoms, believed to be due to an inherited *Gouty Diathesis.* Trial in one or two cases of *inveterate Chronic Gout* has afforded much satisfaction to us, free diuresis being provoked and *followed by relief of symptoms.*"

Uric Acid Diathesis, Gravel, Gout, Eczema, &c.

Dr. J. S. Wellford, *Professor of Diseases of Women and Children, Medical College of Virginia.*

" I have paid a great deal of attention to *Urinary Troubles*, and have frequently and freely prescribed the Lithia Water in their treatment, *with the very best results.* In all the various forms of the *Uric Acid Diathesis*, whether as well-formed *Gravel* or *Gout*, or in the milder forms of *Gouty Dyspepsia* or *Nettlerash* in their varieties, *I know of no Mineral Water which I consider ot all equal to that of Spring No. 2.*

"In many *Skin Diseases* of old age, dependent on the *Uric Acid Diathesis*, such as *Eczema*, &c., this water acts most beneficially."

Chronic Inflammation of the Bladder, whether resulting from Stone, Enlarged Prostate, or Neglected Gonorrhœa.

Dr. Robert Battey, of Georgia, Suggestor of Battey's Operation, Member of the American Medical Association, &c.

"I would state that I have been using the Buffalo Lithia Water, No. 2, in my practice for three years past, in cases of *Chronic Inflammation* of the *Bladder*, whether induced by *Stone*, by *Enlarged Prostate* in the aged, or by *Neglected Gonorrhœa*, and have secured excellent results, which encourages me to prescribe it for the future."

Affections Peculiar to Women. Prophylactic Virtues of the Water.

Dr. Goodrich A. Wilson, Somerset, Granville Co., N. C., Member of North Carolina Medical Association.

"But I should be doing but meagre justice to the merits of the Buffalo Lithia Water, if I failed to refer to its great *Prophylactic* virtues. Hundreds of young *Females* attain to the age of puberty without realizing the healthful performance of their great function. They become *Nervous, Hysterical, Chlorotic.* Hundreds of young men attain to the same age without developing into manhood. They become pale, feeble, irresolute, hypochondriacal, and perhaps the end is *Phthisis*, or an *Insane Asylum*. In all such cases the best results may be anticipated from a season at these Springs."

Diseases of the Uric Acid Diathesis, and Catarrhal Affections of the Bladder.

Dr. E. J. Doering, of Chicago, Illinois, late Surgeon U. S. Marine Hospital Service.

"In my experience, the Buffalo Lithia Water is an efficient remedy in the treatment of diseases depending on the *Uric Acid Diathesis*. I have prescribed it quite extensively in *Catarrhal Affections* of the *Bladder*, with excellent results. I am fully satisfied that it possesses great value as a general remedial agent, and I shall take great pleasure in recommending it to other physicians."

Malarial Cachexia, Atonic Dyspepsia, Affections Peculiar to Women, &c.

Dr. O. F. Manson, of Richmond, Va., Professor of General Pathology and Physiology in the Medical College of Virginia.

"I have observed marked sanative effects from the Buffalo Water in *Malarial Cachexia, Atonic Dyspepsia*, some of the *Peculiar Affections of Women, Anæmia, Hypochondriasis, Cardiac Palpitations*, &c. It has been especially efficacious in *Chronic Intermittent Fever, numerous cases of this character, which had obstinately withstood the usual remedies, having been restored to perfect health in a brief space of time by a sojourn at the Springs.*"

7

Acid Dyspepsia, Gravel, Gout, Rheumatism, Nausea of Pregnancy, Uræmic Poisoning complicating that Condition, Malarial Poisoning, Cutaneous Diseases, &c.

Dr. Wm. H. Doughty, Professor of Materia Medica and Therapeutics, Medical College of Georgia ; Surgeon in charge of General Hospital, Confederate Army ; Member of American Medical Association, &c.

" I prescribe the Buffalo Lithia Water with the utmost confidence in all forms of *Indigestion*, due to *Chronic Catarrh* of the *Mucous Membrane*, with *excess of Acid ;* also in the secondary or symptomatic *Dyspepsia* of *Uterine* and *Renal* origin. It is the only reliable treatment known to me for the permanent relief of *Gravel* and the antecedent conditions that determine it. In *Gout* and *Rheumatism*, I regard it as a remedy of great value. Over the *Nausea* and *Vomiting of Pregnancy*, particularly in the latter months where *Uræmic conditions* are possibly established, and in *Puerperal Convulsions*, *Uræmia* co-existing, it often exerts marked control. In *Hepatic Disorders*, especially where attended with *Jaundice* or *Biliary Calculi*, it is an efficient remedy. In *Genito-Urinary* diseases, especially *Catarrh* of the *Bladder* in *Females*, I have found it very efficacious. Whenever *Menstrual* or *Uterine* disorders are produced or intensified by such a state of the Bladder, it becomes a valuable adjunct to other treatment. I have seen some very gratifying results from this water in *Chronic Cutaneous Diseases*."

Gout.

Dr. John T. Metcalf, of New York, Emeritus Professor of Clinical Medicine, College of Physicians and Surgeons, New York.

" I have for some years prescribed the Buffalo Lithia Water for patients, and used it in my own case for *Gouty* trouble, with decided beneficial results, and I regard it certainly as a very valuable remedy."

Gout, Rheumatic Gout, Rheumatism, Stone in the Bladder, Bright's Disease of the Kidneys, &c.

Dr. Wm. B. Towles, University of Va., Member of Medical Society of Va.

" In *Gout, Rheumatic Gout, Rheumatism, Stone in the Bladder*, and in all Diseases of the *Uric Acid Diathesis*, I know of no remedy at all comparable to Buffalo Lithia Water, Spring No. 2. In a single case of *Bright's Disease of the Kidneys I witnessed very marked beneficial results from its use*, and, from its action in this particular case, I should have great confidence in it in this disease."

Albumen and Suppression of Urine in Scarlet Fever. Chronic Alcoholism.

Dr. G. W. Semple, Hampton, Va., Ex-President Medical Society of Va.

"In *Scarlet Fever* I have known the Buffalo Lithia Water, Spring No. 2, to restore a healthy and abundant secretion of *Urine* when it was highly charged with *Albumen* and the secretion almost suppressed.
" I regard it as exceedingly valuable in the treatment of *Chronic Alcoholism.*
" It not only effectually eliminates *Alcohol* and effete matters, but at the same time meets other important indications.
" It relieves the *Irritable Stomach*, increases the appetite, promotes digestion and assimilation, and gives tone to the shattered *Nervous System*."

Diseases of Uric Acid Diathesis.

Dr. Lewis D. Mason, *of Brooklyn, New York, Professor of Surgery, Long Island Hospital Medical College.*

"I have prescribed the Buffalo Lithia Water with very satisfactory results. In diseases, especially of the *Uric Acid Diathesis,* I regard it as a remedy of great power and excellence."

Chronic Diseases of Females. Affections of the Kidneys and Bladder. Especially Hæmaturia.

W. F. Barr, M.D., LL.D., *Member Medical Society of Virginia ; Ex-President of the Abingdon and of the Southwest Virginia Medical Society ; Corresponding Member of the Boston Gynæcological Society.*

"Experience in its use in a number of cases, enables me to bear testimony to the virtues of the Buffalo Lithia Waters of Mecklenburg county, Virginia, in *Chronic Diseases of Females* and in *Affections* generally of the *Kidneys* and *Bladder.* I have relieved cases of *Hæmaturia* with it that had resisted all remedies used and recommended by Practitioners and Standard Authors."

Diseases of the Uric Acid Diathesis, Affections of the Uterine and Urinary Organs. Sterility in Women, and Impotency in Men, Malarial Poisoning, Indigestion, &c.

Dr. Wm. F. Carrington, *Hot Springs, Ark., Surgeon (retired) U. S. Navy, Surgeon Confederate States Navy, Medical Director Army of Northwestern Virginia.*

"The Buffalo Lithia Water, Spring No. 2, has signally demonstrated its remedial power in *Gout, Rheumatism, Uric Acid Gravel,* and other maladies dependent upon the *Uric Acid Diathesis.*

"It not only eliminates from the blood the deleterious agent before it *crystallizes,* but dissolves it in the form of *Calculi,* at least to a size that renders its passage along the ureters and urethra comparatively easy.

"I am able to attest its beneficial effects in all, and .ts positive curative virtues in many of the *Affections* of the *Mucous Membrane,* as well as *Functions* of the *Urinary* and *Uterine Organs.* Even in the mechanical derangement of *Prolapsus Uteri,* it often relieves by its tonic power, re-establishing the general health ; relieving, where caused by those affections, *Sterility* in women, *Impotency* in men. In *Diabetes Mellitus,* I have no experience in its use, and cannot give the rationale of its action in such cases. There is, however, satisfactory evidence that it is a remedy of decided value in the treatment of this distressing malady. In all *Malarial Poisoning,* it enjoys a deservedly high reputation as a curative agent. It increases the appetite, promotes digestion, neutralizes acid. and is highly beneficial in some forms of *Dyspepsia.*"

Albuminuria of Scarlet Fever.

Extract from Communication of Dr. C. W. P. Brock, of Richmond, Va. (Member Medical Society of Va.), in the Virginia Medical Monthly for November, 1878.

"During the epidemic of *Scarlet Fever,* which has been prevailing for a year in this city, and is yet scarcely ended, I have been giving my patients Buffalo Lithia Water *ad libitum.* and to the exclusion of all water for drinking purposes. In no case since I have pursued this course have I seen even a trace of albumen in the urine of *Scarlatinal* patients, either during the attack or the convalescence. With this experience in a number of cases, and hearing of favorable results in the practice of other physicians in this city who have used the same means, I have thought it worthy of note in your journal. If this result is *proctor hoc*, we are enabled to rid *Scarlet Fever* of one of its most dangerous concomitants."

Urinary Calculi.

Dr. Henry M. Wilson, of Baltimore, Ex-President Medical and Chirurgical Faculty of Maryland.

"My experience in the use of the Buffalo Lithia Springs' Water has not been large, but it is of such a positive character that I do not hesitate to express my preference for it as a *Diuretic* in *Urinary Calculi*, *over all other waters that I have ever used.*"

Gout and Rheumatic Affections.

Dr. Charles B. Nancrede, Surgeon to the Episcopal Hospital, and St. Christopher's Hospital, Philadelphia.

Dr. Nancrede in the "International Encyclopædia of Surgery," edited by Dr. John Ashhurst, Jr., Professor of Clinical Surgery in the University of Pennsylvania, upon the subject of the treatment of *Gouty* and *Rheumatic Affections*, advises among other remedies "the natural *Alkaline* waters," and expresses the opinion that the Buffalo Lithia Water is one of the best *Alkaline* waters of this country, and adds the following:

"The only natural water used at the patient's house, of which I have any experience, is the Buffalo Lithia Water, *which I have used with undoubted advantage in my own person.* It should be taken in large quantities and for a long time."

Gouty Diathesis.

Dr. Jno. R. Page, Prof. Zool., Bot. and Agric., University of Virginia.

"I have observed decided benefit from the use of the Buffalo Lithia Water in *Gout, Lithiasis, Lumbago,* and *Sciatica,* due to the same 'Materies Morbi,' and am fully satisfied of its great value in the treatment of *all affections due to a Gouty Diathesis.*"

Lithiasis, Rheumatism, accompanied by Deposits of the Urates.

Dr. G. W. Pope (Homœopathic), Washington, D. C.

"I have prescribed the Buffalo Lithia Water in cases of *Lithiasis* and *Chronic Rheumatic Affections,* accompanied by *Deposits* of *Urates* of *Soda,* and generally with excellent results. In one case of *Heart Disease,* with *Aortic Obstruction* and consequent regurgitation, caused by *Calcareous Concretions* upon the *Valves*—and which case was associated with a marked *Lithic Acid Diathesis*—the Buffalo Lithia Water, in tumblerful doses, thrice daily for a week or two at a time, had a most happy effect in clearing up the urinary deposits, and placing the patient, in respect both to the frequency and severity of the Cardiac attacks, in a much better condition than formerly. It was given during many attacks, and always with unmistakably good results."

Nausea and Urœmic Poison of Pregnancy.

Dr. *Caleb Winslow, 23 McCulloh Street, Baltimore, Member of the Medical and Chirurgical Faculty of Maryland.*

"I have found the Buffalo Lithia Water, Spring 'No. 2,' of marked service in relieving the *Nausea of Pregnant Women.* I frequently resort to it at intervals during the whole course of *Pregnancy.* Being *antacid, laxative, diuretic and tonic,* it seems well adapted to relieve the disturbances usually attendant upon *Gestation,* and I have no doubt its free use might remove *Urœmic Poison,* and prevent *Convulsions* produced thereby."

Gout, Rheumatism, Rheumatic Gout, Lithic Acid Culculi, &c.

Dr. *Algernon S. Garnett, Hot Springs, Arkansas, Surgeon (retired) U. S. Navy.*

"My experience in the use of Buffalo Lithia Water is limited to the treatment of *Gout, Rheumatism,* and that hybrid disease '*Rheumatic Gout*' (so-called), which is in contradistinction to the *Rheumatoid Arthritis* of *Garrod.*

"I have had excellent results from this water in these affections, both *in my own person,* and in the treatment of patients for whom I have prescribed it. Of course the remedial agent is its contained *Alkalies* and their *solvent* properties.

"Hence, it is a *prophylactic as well as a remedy* in *Nephritic Colic* and forming *Calculi,* when due to a redundancy of *Lithic Acid.*"

Affections Peculiar to Women.

Dr. *J. H. Tucker, Henderson, N. C., Member of Medical Society of North Carolina.*

"Of the special adaptation of the Buffalo Lithia Waters to the *Peculiar Diseases of Females,* I have no question. In *Chloro-anemic* diseases, in *Leucorrhœa,* in some of the varieties of *Dysmenorrhœa,* and in all those functional derangements resulting from *Nervous Atony,* I prescribe these waters *with almost the same confidence that I do quinine in Chills and Fever.*"

REPORTED CASES.

Affections of the Kidneys and Bladder, Stone of the Triple-Phosphate of Ammonia and Magnesia Variety, Four Ounces of Stone passed under the action of this Water, where the Surgeon's Knife had been Ineffectual.

Case of Mr. C———, of North Carolina, stated by Dr. G. Halstead Boyland, Late Professor of Surgery, Baltimore Medical College, Late Surgeon French Army (Decorated), Member of the Medical and Chirurgical Faculty of the State of Maryland, &c.

"The case of Mr. C———, of North Carolina, who arrived at the Springs June 21st, affords undoubted evidence that Buffalo Lithia Water, Spring No. 2, is a *Solvent for Urinary Deposit,* commonly called '*Stone in the Bladder.*' About a year previous he was operated upon for Stone, the operation affording but partial and temporary relief. He complained of pain in the Lumbar Region, and pain and irritability of the neck of the Bladder. He was emaciated; suffering greatly from Insomnia; and his general condition very unfavorable. Upon arrival at the Springs he was passing small quantities of a Urinary Deposit of the *Triple-Phosphate of Ammonia and Magnesia* variety. Large flakes of bloody mucus were found in the urine. For the relief of present suffering he was making frequent and free use of opiates. He was put upon the water of Spring *No.* 2—from six to eight glasses a day. In a few weeks the *Solvent Properties* of the Water were evident in the diminished consistency of the Deposit, the increased quantity discharged, and by its change from *Concrete Lumps to Fine Sand*, which he discharged to the amount of *four ounces*. The quantity, however, diminished, until, after a stay of eight weeks at the Springs, he has returned home with the *Deposit* dissolved and washed out of the system, and the Diathesis, *Fons et origo morbi*, altered. There has been a disappearance of the attending distressing symptoms described, and great improvement in his general condition."

Stone in the Bladder (Uric Acid) " *Destroyed by the action of the Water, by means of Solution or Disintegration.*"

Case of Dr. B. J. Weistling, Middletown, Pa.; stated by himself.

" Experience in its use in *Stone of the Bladder* in my own person enables me to attest the efficacy of the Buffalo Lithia Water in this painful malady. After having been long subjected to sufferings, the intensity of which cannot be described, I have, under the influence of this water, passed an ounce of *Calculi* (*Uric Acid*), some of which weighed as much as four grains, affording inexpressible relief and leaving me in a condition of comparative ease and comfort.

" On one occasion I passed thirty-five *Calculi* in forty-eight hours. The appearance of this *Calculus Nuclei* indicates unmistakably, I think, that they were all *component particles of one large Calculus, destroyed by the action of the water, by means of solution and disintegration.* At my advanced period of life (I am seventy-seven years and six months of age), and in my feeble general health a surgical operation was not to be thought of, and the water seems to have accomplished all that such an operation, if successful, could have done. Besides greatly increasing the quantity of the Urine, this water *exerts a decided influence on its chemical constitution*, rendering it *rapidly neutral, if previously acid*, and afterwards *alkaline;* from being *high-colored* it becomes *pale*, and having *deposited copiously* it becomes *limpid* and *transparent*."

Uric Acid Calculi

Case of Dr. J. D. Eggleston, Hampden Sidney, Va., Member Medical Society of Virginia; stated by himself.

"The action of the Buffalo Lithia Water in the so-called *Uric Acid Diathesis* is especially happy.

" I had a very gratifying experience in a case of Calculi in the Bladder, of Uric Acid variety, in my own person. Between February 1st and May 1st, 1881, I had thirty attacks of great severity, passing in each one of them a number of Calculi, aggregating 165. Finding no permanent benefit from any treatment of the profession, I made a visit to the Buffalo Lithia Springs. The happy adaptation of this water to my condition was soon evident, and its use for a few weeks arrested the *Calculus* formation and resulted in my entire relief, which I have reason to hope is permanent, as more than a year has elapsed since there has been any manifestation of the trouble."

Stone in the Bladder (Lithic Acid), and Hemorrhage from the Bladder.

Case of Mrs. ——, stated by Dr. J. B. Jones, Charlotte, N. C., Member Medical Society of North Carolina.

"Mrs. —— suffered with *Stone in the Bladder*, composed of alternate concentric layers of *Calcareous matter* and *Lithic Acid*, attended with occasional alarming *Hemorrhage* from the Bladder.

"In a critical condition I prescribed for her Buffalo Lithia Water, Spring No. 2, the continuous use of which *arrested the formation of Stone*, and diminished both the frequency and the violence of the *Hemorrhage*.

"I have frequently prescribed this water in chronic affections of the *Kidneys* and *Bladder* with the best results, and in such cases I regard it a remedy of greater potency than any mineral water of which I have any knowledge. In all irritable conditions of the *Urinary Passages* it will be found highly beneficial."

Stone in the Bladder (Uric Acid), Symptoms of Bright's Disease.

Case of Mrs. ——, stated by Dr. David E. Smith, of Bronxville, Westchester, N. Y.

"Mrs. ——, was subject to severe attacks of *Gout*, a consequence of an inherited *Gouty Diathesis*, followed after some time by *Stone in the Bladder*. The limbs were very *Œdematous*, so much so as to pit readily on pressure with the finger, leaving an indentation long after its removal. The Urine was loaded with *Urates* and *25 per cent Albumen*, and the microscope revealed *Casts*, showing unmistakably, as I thought, *Bright's Disease of the Kidneys*. I prescribed the Buffalo Lithia Water, Spring No. 2, which afforded prompt relief in the *Gouty Affection*, and resulted, in a few weeks, in the passage of a *Stone ⅜* of an inch long by ¼ of an inch in diameter. Under the continued use of the water the Urine has been relieved of *Albuminous Impregnations*, and restored to a normal condition, no *Casts* can be discovered, the *Œdematous* condition of the limbs has been relieved and the general health of the patient, to a great extent, restored."

Stone in the Bladder of the Triple-Phosphate of Ammonia and Magnesia Variety.

Case of Mr. B ——, of Fulton, Ky., stated by Dr. James Shelton, residing near the Buffalo Lithia Springs, Va.

"The case of Mr. B——, of Fulton, Ky., a recent visitor at the Buffalo Lithia Springs, furnishes, I think, satisfactory evidence of the *Solvent* power of the water of Spring No. 2 in certain varieties of *Stone in the Bladder*.

"He informed me that he had been for a number of years subject to attacks from which he was occasionally relieved by the passage of *Stone*, some of them weighing as much as six grains. It was evident that his general health was greatly impaired. Several years previous to his visit to the Springs he had used this water at home, and as the result of its action discharged a number of small *Calculi*. Notwithstanding this fact, however, there was about this time such aggravation of his general condition that by the advice of his medical attendant he abandoned the use of the water.

"It subsequently proved, however, that its action had been most beneficial, as he enjoyed long afterwards immunity from these attacks. The protracted use of the water at the Springs was followed for some time by a free discharge of small *Calculi*, the quantity exceeding *three ounces*, which proved to be of the *Triple-Phosphate of Ammonia and Magnesia Variety*.

"Before his departure this *Calculus* discharge ceased, and he left the Springs apparently free from disease, and with his general health restored, having especially gained largely in flesh."

Uric Acid Gravel.

A Case stated by Dr. Hasford Walker, of Georgetown, S. C.

"Mr. George C. Congdon, of this place, suffering from *Uric Acid Gravel*, has been under my treatment for some time past. He had as many as six separate attacks (all of them of very great severity), between June the 28th and July the 16th, a space of eighteen days. He declined in flesh and strength rapidly, and to such an extent as to excite serious apprehensions as to the result. Having exhausted the most approved remedies of *Materia Medica* without beneficial results, I prescribed for him, as a last resort, Buffalo Lithia Water, Spring No. 2, with the happiest effects. In a very short time after beginning its use, he passed a *Calculus* of about six grains in weight, has not been troubled since, and is now, to all appearance, entirely well."

Stone in the Bladder, Phosphatic.

A Case Stated by Dr. B. S. Hardy, Hookerton. Green Co., N. C.

"My son, a young man, suffered from *Stone* in the *Bladder*, of a *mixed character*, consisting chiefly of *Calcium, Carbonate* and *Phosphate.* After persistent use of all remedies indicated in the case without benefit, I put him upon Buffalo Lithia Water, Spring No. 2, the use of which, for some six weeks, resulted in the passage, at short intervals, of *Gravel*, of small size, and at times of particles of *Sand*, followed, some time afterwards, by the discharge of a *Stone* weighing twelve grains (the largest, I am confident, I ever knew to pass through the Urethra), virtually ending his troubles, since which time any unpleasant symptom has been relieved by the use of the water for a short time. Its action in this case has been indeed wonderful."

Bright's Disease of the Kidneys, Uræmic Poisoning, &c.

A Case stated by Dr. Jno. W. Williamson, Boydton, Va.

"Several years since, when a resident of the State of Tennessee, my wife suffered from well-defined *Bright's Disease* of the *Kidneys*, resulting in *Uræmic Poisoning.* After a signal failure of every remedy suggested by several eminent medical men, and when her condition was regarded as well nigh hopeless, trial was made of Buffalo Lithia Water, Spring No. 2. The result was relief from the threatening symptoms so prompt and decided as to be almost incredible to any but an eye-witness.

"She continued the use of the Water for several months, making a complete recovery, having no return of the malady, and is now in good health.

"I will add that in diseases of this character *I know of no remedy, either among mineral waters or medicines, at all comparable to this Water.*"

Bright's Disease of the Kidneys.

Case of Mr. ——, stated by Dr. Z. M. Puschall, Oxford, N. C., Member Medical Society of North Carolina.

"I spent the summer of 1880 at the Buffalo Lithia Springs, and while there witnessed the marked beneficial action of this water in a case of *Bright's Disease of the Kidneys.* Mr. ——, the sufferer, reached the Springs in a condition of emaciation and extreme exhaustion. The *Urine* was heavily charged with *Albumen*, and its specific gravity decidedly below the healthy standard, with general

Œdema and *Comatose* symptoms. In three weeks there was evident improvement, which continued during a stay of two months at the Springs, the *Urine* becoming free from *Albumen*, natural both in appearance and quantity, and regaining to a great extent a healthy density and a disappearance of the *Dropsical* symptoms. In the meantime there was great improvement in the general health, the patient gaining some twenty-five pounds in weight, and leaving the Springs in a comfortable condition."

Hæmaturia or Hemorrhage from the Bladder.

Dr. E. M. Campbell, of Abingdon, Va., Member of the Medical Society of Virginia.

"The most aggravated case of *Hæmaturia* which I ever saw was relieved by the Buffalo Lithia Water, after an entire failure of the treatment usual in such cases.

"In affections, generally, of the *Kidneys* and *Bladder*, I must regard it as a remedy of great value."

GOUT, RHEUMATIC GOUT, &c.

Gout.

Case of Dr. P. W. Young, of Oxford, N. C., Member of the State Medical Association; Surgeon in charge of General Hospital, Confederate States Army ; stated by himself.

"In the month of August, 1879, prostrated by a severe attack of *Gout*, with which I had then suffered for several months, *and in such a condition that I scarcely hoped ever to be able to resume active professional life,* I visited the Buffalo Lithia Springs. I was without appetite; my digestion imperfect and painful; my *Nervous System* shattered, and the victim of *Neuralgia* in all its forms; my spirits greatly depressed and so feeble that I could walk but a short distance without assistance. The first noticeable effect of the water was upon the secretion from my kidneys and skin, both of which were acting very imperfectly and failing properly to depurate the blood. The secretion from the kidneys had been for months scanty, high-colored, of high specific gravity, very acid in its reaction, and depositing, upon standing, a very large sediment. The action of the water was apparent in forty-eight hours, the secretion becoming more abundant, less highly-colored, and depositing less sediment. With the re-establishment of a healthy state of these two important secretions, the great depurators of the blood, my convalescence was soon and rapid; my appetite returned, my digestion became perfect, and my *Nervous System* regained its tone. I became more cheerful and hopeful, and gradually regained my strength. My general health is now perfect, and I have resumed the active practice of my profession."

Case of Dr. J. A. Hanby, of Patrick C. H., Va. ; stated by himself.

Rheumatic Gout.

"For four years I was afflicted with *Rheumatic Gout* to an extent which incapacitated me entirely for the discharge of the duties of my profession, and was finally reduced to such a condition as to subject me for the most part to confinement to my bed. By the advice of one of my medical attendants and emphatically as a *dernier resort*, I determined to make use of the Buffalo Lithia Water, Spring No. 2, I am frank to say, without faith in its virtues, having but little confidence in mineral waters. The use, however, of a few cases of the water was

15

attended by beneficial results, so remarkable, that I was soon able to be out of bed and upon my feet, and my improvement has continued until I am now actively engaged in the practice of my profession, meeting, without any unusual inconvenience, all the exposure and hardship incident to the life of a physician in a mountain country. I cannot, in candor, do otherwise than ascribe my recovery solely to this water, the value of which I regard as beyond estimation.''

DYSPEPSIA.

Chronic Gastric Catarrh and Uric Acid Calculi.

A Case stated by Dr. Jno. C. Coleman, of Scottsburg, Va., a retired Surgeon of the U. S. Navy.

"Mr. C. was for a number of years a sufferer from *Chronic Gastric Catarrh.* While his diet was exclusively *tea* and *crackers, bread* and *milk* and other similar articles, it was frequently thrown off in an undigested state soon after taking it, and at times he discharged from an empty stomach a strongly acid glairy mucus. A marked *Uric Acid Diathesis* supervened, consequent upon which he suffered for a period of some two years great *Vesical Irritation* and possible *Cystitis,* attended by pain so intense and constantly present, as to require that he should be kept for the most part under the influence of opiates. After a persistent, but ineffectual exhibition of all remedies supposed to be indicated in the case, he was put for the latter affection upon the Buffalo Lithia Water, Spring No. 2, with the happiest possible effect.

" In a few weeks, after commencing the use of it, the irritable condition of the *Bladder* was so far relieved that he was enabled to dispense entirely with the use of opiates. At the expiration of some eight weeks he had an attack of unusual severity, from which he was relieved by the discharge of a *Calculus,* followed at short intervals by the discharge of three others, which proved to be the termination of this trouble, as from that time there was entire subsidence of the painful symptoms described, and the *Bladder* resumed its natural state.

" While prescribed with special reference to the relief of the *Irritable Bladder,* the action of the water was not less surprisingly happy in the *Gastric Affection,* with remarkable promptness correcting the highly acid condition of the *Stomach,* restoring a healthy digestion and assimilation, and *tone* and *vigor* to the depressed *Nervous System.*

" In a few months he was able to eat, with perfect impunity, the coarsest articles of diet. He is now, after a lapse of several years, in robust health, having had no return of these painful maladies.''

Chronic Dyspepsia with Spasmodic Gastralgia, Extreme Irritability of the Stomach, &c.

Case of Jno. P. Keeling, Esq., stated by Dr. S. S. Keeling, Norfolk, Va., Member Medical Society of Virginia.

' Mr. Jno. P. Keeling labored under *Chronic Dyspepsia,* and was always subject to violent attacks of *Spasmodic Gastralgia* immediately upon taking food into the stomach, which attacks were not at all amenable to treatment. Not unfrequently the stomach rejected everything in the way of food or drink, and he was of necessity confined to the lightest possible articles of diet, meat and vegetables being entirely excluded. He became so prostrated that it was with difficulty he could walk across his chamber floor, and had often to be lifted about. Such was the state of his *Nervous System that great solicitude was felt as to his mental condition.*

"He visited Baltimore for medical aid, and was for many months under the treatment of some of the most eminent men of the profession in that city, but without beneficial result, and was finally advised that he had nothing to hope from remedies.

"Returning in an extremely critical condition to his home in the county of Princess Anne, he came under my professional care. Satisfied that medicine was unavailing in the case, I advised the Buffalo Lithia Water, Spring No. 2. His stomach, however, was in a highly irritable condition, and I found it necessary to administer it in very small quantities; and it was at first given not exceeding an ounce at a dose, repeated at stated intervals. At the expiration of the third day the irritability of the stomach was decidedly less, and the quantity was then increased from day to day until the thirteenth day, when I found that the patient could take twelve ounces, which I regard as a maximum dose at any time. Persisting in its use, on the twenty-eighth day he was free from pain, the stomach in a normal condition, readily receiving both solids and liquids in moderate quantities, strength greatly increased, and nervous symptoms entirely relieved. At the expiration of the seventh week he was able to attend actively and regularly to his business upon the farm. His recovery, which I regard as one of the most remarkable I ever knew of, I attribute entirely to the Buffalo Lithia Water."

AFFECTIONS PECULIAR TO WOMEN.*

Suppression of the Menstrual Flow, Suppression of Urine, Hæmaturia, &c.

Case of Mrs.——, stated by Dr. John T. Winfield, of Broadway Rockingham County, Virginia

"I was called professionally to see Mrs. ———, and found her complaining of excruciating pain in the head, back and limbs, attended with nausea and vomiting. Her kidneys were torpid, and the urine—which had to be drawn by means of the catheter—scanty and charged with blood. She had passed *eight months without experiencing the Menstrual Flow.* * * * During an extreme illness of many weeks, every remedy that seemed at all indicated in the case was faithfully tested, but without result, save an occasional slight alleviation of the distressing symptoms. The kidneys remained inactive, the average daily amount of fluid drawn from the bladder did not exceed two ounces, and more than half that quantity was blood. *On one occasion four days passed without any secretion at all from the Kidneys.* It became evident that the patient must die unless some remedy could be found of much greater efficacy than any that had been tried. In this extremity it was determined to make an experiment in the case with Buffalo Lithia Water, a case of which was ordered with but a faint hope of keeping her alive until its arrival, by making the skin and bowels perform the eliminating functions of the kidneys. So soon as the water was received, the patient was put upon it to the exclusion of all other treatment. The first day's use of the water showed marked improvement in the action of the kidneys. The urine daily increased in quantity, while the proportion of blood in it rapidly diminished, and other distressing symptoms soon disappeared. At the end of the week the catheter was dispensed with, and the urine was voided naturally, with but a trace of blood, and in twenty days the *Menstrual Flow* appeared, and was followed by the early convalescence and perfect restoration of the patient."

Dysmenorrhœa, Chronic Inflammation of the Neck of the Womb, Leucorrhœa, &c.

Case of Mrs. A——, stated by Dr. R. D. DeL. French, of Washington City, D. C.

"Mrs. A——, a lady of delicate physique and highly sensitive nervous organization, suffered for a number of ears from *painful Menstruation, Inflammation of*

17

the *Womb* and profuse *Leucorrhœa.* Three months ago I prescribed for her the Buffalo Lithia Water, and I am happy to be able to state that it has accomplished what protracted medical treatment had failed to do, and that her troubles have entirely disappeared, and that she is now in the enjoyment of better health than she has had for years. This was a case, in view of all its complications, of no ordinary kind; consequently the results obtained are especially worthy of attentive consideration. From the action of the water in this case, which had so long resisted the most approved remedies, I can but regard it as one of the most potent of known agents in functional derangements of the Uterus, especially those of a chronic character."

Phrenitis, Suppression of the Catamenia, &c.

A case stated by Dr. Joseph S. Edie, of Christiansburg, Va., Member Medical Society of Virginia.

" In June last I had a case of *Phrenitis* in a young female. After the furious delirium and fever had passed off she remained melancholy, with greatly impaired memory, suppression of the *Catamenia,* and deficient action of the kidneys. She improved gradually, but there was not an entire restoration of the mind. To my surprise, the suppression of the *Catamenia* continued. The remedies which I thought indicated in the case proving ineffectual, I put her on the Buffalo Lithia Water, of Spring No. 2. In a few weeks after she commenced the use of it, the suppressed secretion was restored, which was soon followed by the restoration of her memory, spirits and general health. I ascribe the recovery to the action of the water."

Irregular and Painful Menstruation.

Case of Miss——, stated by Dr. E. Jones Williams, No. 17 Patuxent St., Baltimore.

" Miss——, twenty years of age, informed me that her *Menstruation* had always been irregular, appearing generally twice and sometimes three times a month, and that during these periods her sufferings were intense. She complained of a want of appetite, imperfect and painful digestion, and of sleepless nights. She was emaciated and laboring evidently under great mental depression. She had been treated at different times, but without effect, by eminent members of the profession. I prescribed for her the Buffalo Lithia Water, Spring No. 2, the use of which for some weeks restored the *Menstrual* function to a normal condition, resulting in the restoration of her general health."

* See statement of the late Dr. J. Marion Sims, the world-renowned specialist in *Diseases of Women.* His general statement that he had used this water in his practice and found it in many cases "HIGHLY EFFICACIOUS," must of course be understood as referring to the use of it in cases coming under the head of his Specialty, Diseases of Women.

DISEASES OF PREGNANCY.

Suppression of Urine, and Threatened Convulsions in Pregnancy. Value of this Water as an Alkaline Diuretic. Acid Dyspepsia.

Dr. Preston Roane, of Winston, N. C., Member of the State Medical Association.

" In a case of almost total *Suppression of Urine,* in a woman in the latter stages of gestation, with strong threatenings of convulsions, after exhausting, without effect, the most potent Diuretics of Materia Medica, I put her upon the Buffalo Lithia Water, Spring No. 2, half a gallon a day, which produced a copious action on the *Kidneys,* and was followed by relief of the alarming symptoms. I attributed the safe termination of the case entirely to the use of this water. It unquestionably possesses extraordinary potency and value as an *Alkaline Diuretic.* I have found it an efficient remedy in *Acid Dyspepsia.*"

Gastric Irritability Superinduced by Pregnancy.

Dr. Roy B. Scott, Trinity Mills, Dallas County, Texas.

"A lady in three pregnancies suffered from extreme *Gastric Irritability*, so much so in two of them as to result in the birth of seven-month children, in the third the symptoms were much more aggravated, and such as to excite serious apprehension as to the life of the patient.

"After using, without a palliation of her sufferings, all the well-recognized agents of Materia Medica, I put her upon the Buffalo Lithia Water. Spring No. 2, which afforded prompt and permanent relief, its continued use enabling her to carry the foetus to its full time."

URÆMIC POISONING.

Uræmia.

Case of Miss C——, stated by Dr. B. F. Hopkins, of the Warm Springs, Va., Member Medical Society of Virginia.

"Miss C——, twenty years of age, was prostrated by a severe attack of *Typhoid Fever*, which was followed by *Uræmia* developing itself at the expiration of two months. The attendant symptoms were such as to excite serious apprehension and proved wholly unamenable to the treatment indicated in the case; the patient grew gradually worse, until I regarded her condition as hopeless. At this time a friend of the young lady suggested the Buffalo Lithia Water. Satisfied that no injury could result from its use, I gladly adopted the suggestion. The result was matter of equal gratification and astonishment. Under the influence of the water the *Kidneys* promptly resumed a healthy action, the *drowsiness* disappeared, and in two weeks the patient was up and walking about the house, which she had not been able to do before for several months, and her improvement continued until she was in usual health."

BILIARY CALCULI, OR GALL STONES.

Biliary Calculi.

Case of Mrs. ——, stated by E. C. Laird, Haw River, Alamance County, N. C., recently of Virginia.

"Mrs. C—— was, for a number of years, subject to frequent attacks of *Biliary Calculi*, continuing, at times, for several days, and at others, for a much longer period. The paroxysms of pain were of unusual severity, and attended by violent nausea and vomiting. These frequent attacks greatly impaired the general health, and more especially the nervous system. In the hope of finding relief, she made several visits to the most celebrated mineral springs of Virginia and North Carolina, and on several occasions visited the Northern cities for medical treatment. All, however, proved unavailing, and these attacks continued, making their steady inroad upon her already much enfeebled constitution, until two summers ago, when she was sent to the Buffalo Lithia Springs, of Virginia, where she remained for six weeks, using the water with the happiest possible results. Her disease was arrested, there is every reason to believe, permanently, as she has since had no return of it, and is wonderfully built up in general health."

19

Paralysis, resulting from Rheumatic Gout, Dropsical Effusion, &c.

A Case stated by Dr. James Beale, of Richmond, Va., Member Medical Society of Virginia.

"In my own family the use of the Buffalo Lithia Water (Spring No. 2) has exercised the most happy influence. Mrs. Beale commenced the use of this water after a confinement of eighteen months to her room from attacks of *Rheumatic Gout*, which had brought on a *Paralysis*, incomplete, of the lower extremities and of her right hand. Latterly, this condition of things was succeeded by *Dropsical Effusion* in both limbs, rendering locomotion impracticable without assistance. Since using the water, which she has done for several months at home, the *Dropsical Effusion* has disappeared. She walks without assistance and writes legibly, previously having been compelled to employ an amanuensis."

BLOOD POISONING.

Uric Acid Eczema.

A Case stated by Dr. M. M. Jordan, of Boydton, Virginia, Member Med. Soc. Va.

"For a period of some five years I was frequently called to a patient, a gentleman suffering from a most harassing *Eczema*, evidently dependent upon the *Uric Acid Diathesis*. His sufferings were intense, and at times such as to confine him for weeks, and at others, for months, to his bed. In his long ill-health I repeatedly administered the remedies advised by the best authorities in such cases, but without any permanent good effect. He was at several different times under the treatment of some of the most eminent of the profession in the Northern cities, spent a summer among the mineral springs of the mountains of Virginia, and a winter at the Hot Springs of Arkansas, but all without finding any decided alleviation of his suffering. The summer following his visit to the Hot Springs, he spent at the Buffalo Lithia Springs, finding in the water of Spring No. 2 the relief which he had in vain sought from so many different sources. Three years have now passed without any return of the disease to such an extent as to occasion serious annoyance, and when there has been any manifestation of it, it has been relieved by the free use of the water."

Septicæmia Blood Poisoning, following Puerperal Fever.

Statement of Dr. B. H. Hite, Hollydale, Lunenburg County, Virginia, Member Medical Society of Virginia.

"Mrs. ——, from exposure to cold and other imprudences, after a violent attack of *Puerperal Fever*, relapsed, her disease assuming a chronic form of irritation of the *Genito-Urinary Organs*, with functional derangement of the *Liver* and *Digestive* system—*Gastric Irritability*, with *Nausea, Vomiting* and *Obstinate Constipation*—large abscesses in the axilla, and one in the lower extremities—pulse frequent, skin dry, tongue smooth and red—extreme emaciation from failure of digestion, nutrition and suppurative drain—her anxious face, colliquative sweats, and other exhaustive symptoms, clearly indicated a hectic tendency and fatal termination. After months of suffering she was put upon the Buffalo Lithia Water with marked and decided improvement, and a gradual restoration to perfect health."

MALARIAL POISONING.

Liver Affection, Jaundice, and Dropsical Effusions, the Sequelæ of Intermittent Fever.

Case of Dr. David G. Smith, Oakley, Mecklenburg Co., Va.

"Some years ago, when a resident of a *Malarious* Southern climate, my general health completely succumbed to oft-repeated and obstinate attacks of *Intermittent Fever*. My whole system seemed to be under the influence of *Malarial Poisoning*. My *Liver* and *Spleen* were involved; I became *Jaundiced*, *Dyspeptic Emaciated* and *Dropsical*. In an exceedingly prostrate and critical condition, I made. a visit to the Buffalo Springs. The action of the water upon my diseased system was prompt and powerful. My *Liver* soon resumed a healthy action; my *Skin* cleared up; my *Digestion* was restored; my *Dropsical* symptoms disappeared and I was in a few weeks a well man. While at the Springs I used no remedies, and the gratifying result stated was of course due entirely to the action of the water."

Intermittent Fever.

A Case stated by Dr. F. J. Gregory, Keysville, Charlotte Co., Va.

"Mrs. —— was for several years subject to frequent recurring attacks of *Intermittent Fever*. While interrupted by appropriate treatment, it seemed impossible to eradicate the *Chill Habit*. As a sequelæ, there was very serious *Functional Disturbance* of the *Liver*, *Splenic Enlargement*, *Œdema* of the lower extremities, and such general depression of the vital powers as to excite serious apprehension for the life of the patient.

"In this condition I advised the Buffalo Lithia Water, Spring No. 2, which proved promptly beneficial, the use of it, for some eight or ten weeks, resulting in the relief of the attacks of *Chill* and *Fever*, and in a remarkable restoration of the general health."

DROPSY.

Case of Mr. L., stated by Dr. James North, No. 2018 N. Twelfth St., Philadelphia.

[Extract from Letter dated Philadelphia, September 24th, 1877, referring to several cases relieved by these waters, which had attracted the attention of Dr. North, on a then recent visit to the Springs.]

"Mr. L., an old gentleman, who appeared to be upwards of seventy years of age, a case of dropsy; was extremely feeble and much swollen; his feet so much so that he could not wear his shoes. Under the influence of the waters the effusion rapidly subsided, and his limbs were reduced to their natural size. Before leaving the Springs he was walking actively about the grounds, and in good health for one of his years."

INDICATIONS FOR, AND MANNER OF, USING THIS WATER.

This water is indicated in all of that large class of disorders dependent upon a *Uric Acid Diathesis, Gout, Rheumatic Gout, Rheumatism, Gravel* and *Stone* of the *Bladder*, in *Chronic Bright's Disease*, and in all affections of the *Kidneys, Bladder* and *Urethra*, requiring *Alkaline* treatment. In the various diseases of the *Digestive Organs*, including *Dyspepsia, Liver Disease, Jaundice, &c.*, it has proved highly efficacious.

It has been successfully employed in the *Summer Diarrhœa of Children*.

In *Habitual Constipation* and in *Hemorrhoids* dependent upon a torpid condition of the *Liver*, it is a remedy of great excellence. The relief afforded by it in *Constipation* is not due to any *decided laxative* property, but to its *alterative power over the secretions*, and while in some of these cases its good effects are prompt, in others the continuous use of the water for weeks or possibly months may be required to accomplish the desired result.

It is an efficient remedy in some of the *Affections Peculiar* to *Women* and especially so in *Suppressed* and in *Difficult* and *Painful Menstruation*.

In *Chronic Malarious Fevers* of every variety, it is confidently claimed to be a specific. In *Blood Poisoning* it will be found a potent agent. It has been found an efficient remedy in *Chronic Gonorrhœa, Gleet,* and *Scrofulous Affections*.

Some distinguished medical men report very remarkable results from its use in the *Albuminuria, Nausea* and *Uræmic Poisoning* of *Pregnancy*, and in the *Albuminuria* and *Dropsical Effusion* of *Scarlatinal* patients.

Distinguished medical authority also ascribes to it a peculiar power as a *Nerve Tonic*, and compares its action as a *tonic* and *alterative* to that of the " *Hypophosphites* of *Lime* and *Soda*."

The best evidence perhaps that it possesses this peculiar power as a *Nervous Tonic* is the remarkable relief which it has afforded in some cases of *Paralysis*, in which its good effects can be ascribed only to its action upon the *Nervous System*.

One or two goblets, of the ordinary size, of this water taken an hour before meals (and where more than one is taken, they should be taken at intervals of a half hour), increases the production of the acid gastric juice, promotes both the appetite and the digestion.

A goblet taken *during* or *after meals* prevents or arrests the fermentations which in certain morbid conditions of the *stomach* take place in the foods in the process of digestion. It is not of course suggested that the same individual should take the water *at, during* and *after meals*, but simply indicate several appropriate times for taking it, to be selected between, according to circumstances and individual experience.

Many persons who cannot ordinarily use *milk* or *cream will find that by adding from a fourth to a third part of the water, they are made not only acceptable but valuable articles of diet*. It has been found useful in the preparation of *artificial food* for infants, as a preventive of *Colic* and *intestinal* derangement, *adding the water to milk until it loses its acidity and becomes neutral or alkaline*. For this purpose a decided preference is expressed for it over lime-water, for the reason that the long-continued use of the latter is regarded as hurtful to digestion

(Continued on 4th page of cover.)

Its Purgative Action.

This water is not, as a general rule, a purgative water. In exceptional cases, however, it acts freely upon the *Bowels* at first, but generally corrects itself in a few days.

Where a cathartic effect is desired, it may generally be obtained by the addition of a teaspoonful of common salt to a goblet of the water in the morning before breakfast.

An Antidote to the Acids of Wines and Liquors.

Many critical judges in such matters regard it as the most efficient of known *Antidotes* to the pernicious acids of *Wines* and *Liquors*, and consequently the most efficient preventive of *Gout, Rheumatism,* and other evils frequently resulting from them.

Suggestions as to Quantity of Water to be Taken, &c.

From six to eight glasses, of the ordinary size, is about the average quantity of this water taken by invalids a day, and this is supposed to be sufficient in most cases; some, however, take from ten to twelve; others only three or four. In derangements of the *Digestive Organs* especially, *the lesser quantity* has often been productive of great good. In *Gout, Rheumatism, Stone, Gravel, or Eczema,* it should be taken freely.

It is impossible to prescribe a standard that would be applicable to all cases. Experience will, perhaps, prove the best guide in the matter. Nor can any general rule be laid down as to the length of time for which the water should be used. While in some cases it is prompt in its action, in others it requires *continuous use for weeks and sometimes for months* to make any decided impression upon the system.

Purchasers Cautioned Against Fraud.

SINCE THE BUFFALO WATERS HAVE ATTAINED SO GREAT REPUTATION, ALL SPRINGS IN VIRGINIA, AT LEAST, ARE CLAIMED TO BE LITHIA WATERS, AND DEVICES, INIQUITOUS AND INNUMERABLE, ARE SYSTEMATICALLY RESORTED TO TO IMPOSE THEM UPON THE PUBLIC. SOME OF THESE WATERS ARE EXTENSIVELY ADVERTISED, AND, UPON THE SAME PRINCIPLE THAT THE THIEF CRIES OUT, "CATCH THE THIEF!" APPEND TO THEIR ADVERTISEMENTS A WARNING AGAINST SPURIOUS LITHIA WATERS, AND THE PROPRIETORS OF ONE OF THEM, TO MAKE DECEPTION MORE ASSURED, ADVERTISES LITHIA WATER, SPRING No. 2, with the object, evidently, of confounding it in the public mind with BUFFALO LITHIA WATER, SPRING No. 2.

THE WATER FOR SALE.

This water is put up in Cases of One Dozen Half-Gallon Bottles, and in no other form.

It is delivered at the Scottsburg Depot of the Richmond and Danville Railroad, or the Clarksville Depot of the Richmond and Mecklenburg Railroad, at $5.00 per Case. ☞ A Discount to Dealers.

All orders for the Water must be accompanied by the money or its equivalent. In future this requirement will be rigidly adhered to in all cases.

THOMAS F. GOODE, Proprietor,

Buffalo Lithia Springs,

VIRGINIA.